CIVIL SURVEYING PRACTICE EXAM

for the California Special Civil Engineer Examination

James R. Monroe, Jr., PE

Professional Publications, Inc. • Belmont, CA

How to Get Online Updates for This Book

I wish I could claim that this book is 100% perfect, but 25 years of publishing have taught me that engineering textbooks seldom are. Even if you only took one engineering course in college, you are familiar with the issue of mistakes in textbooks.

I am inviting you to log on to Professional Publications' website at **www.ppi2pass.com** to obtain a current listing of known errata in this book. From the website home page, click on "Errata." Every significant known update to this book will be listed as fast as we can say "HTML." Suggestions from readers (such as yourself) will be added as they are received, so check in regularly.

PPI and I have gone to great lengths to ensure that we have brought you a high-quality book. Now, we want to provide you with high-quality after-publication support. Please visit us at **www.ppi2pass.com**.

Michael R. Lindeburg, PE
Publisher, Professional Publications, Inc.

Civil Surveying Practice Exam
for the California Special Civil Engineer Examination

Copyright © 1999 by Professional Publications, Inc. All rights reserved. No part of this publication may be reproduced, stored in a retrieval system, or transmitted, in any form or by any means, electronic, mechanical, photocopying, recording, or otherwise, without the prior written permission of the publisher.

Printed in the United States of America

Professional Publications, Inc.
1250 Fifth Avenue, Belmont, CA 94002
(650) 593-9119
www.ppi2pass.com

Current printing of this edition: 3

Library of Congress Cataloging-in-Publication Data
Monroe, James R., 1959–
    Civil surveying practice exam for the California Special Civil
Engineer Examination / James R. Monroe, Jr.
        p.   cm.
    ISBN 1-888577-31-2 (pbk.)
    1. Surveying--California--Examinations, questions, etc.
2. Surveyors--Certification--California--Study guides. I. Title.
TA537.M66   1999
526.9'076--dc21   98-32101
                                                          CIP

Table of Contents

Introduction

California and Guam are currently the only state and territory that include taking the California Special Civil Engineering Surveying Examination as part of their requirements for registration as a licensed civil engineer.

You should begin your preparation by reviewing a wide range of textbooks in this field. You need a solid understanding of the fundamentals and principles of engineering surveying. As others have found, excellent results come from studying *Surveying Principles for Civil Engineers*, by Paul A. Cuomo, PLS, and *1001 Solved Surveying Fundamentals Problems*, by Jan Van Sickle, PLS; both books are published by Professional Publications, Inc. This book is designed to complement the other two publications.

Civil Surveying Practice Exam for the California Special Civil Engineer Examination acquaints you with the board-adopted test plan. The intent of the sample exam is to measure your performance. With it, you can appraise your knowledge and skills before the official examination. Solutions are presented with explanations of the relevant key points and essential steps for solving problems.

After taking the sample exam, you will be able to better identify your strengths and weaknesses in all areas of the test plan, which will help you make an informed decision regarding further review preparation for the Special Civil Engineering Surveying Examination. When you concentrate and work on your weak areas, you will be far better prepared and, as a result, you will improve your performance on the official examination.

The learning style and approach of each examinee to understanding the examination subject matter is unique. For exam preparation, you may do self-study or take formal instruction, whichever you feel suits you best. In addition, taking the sample practice exam in a simulated examination situation with constraints similar to those in the official examination can measure your level of readiness. There is no better way to increase your proficiency in and knowledge of all surveying content areas of the test plan and to improve your chances of passing the examination than to test yourself on the sample exam.

Nature of the Exam

The Special Civil Engineering Surveying Examination is given twice a year, in April and October. It tests the entry-level competency of a candidate to practice civil engineering within the acceptable standards of the profession to protect the public. This exam is open book, and it is administered over a two and a half hour period. The exam contains approximately 52 multiple-choice problems derived from content areas as outlined in the Board-adopted test plan. You are required to answer all the multiple-choice problems. For each question, you will be asked to select the best answer from four choices. You need to record your answers on a machine-scorable answer sheet that will be provided to you when the examination is administered. Your calculations should be performed in the official test book, and they will not be credited or scored. Also, answers marked in the official test book will not be scored, and additional time will not be permitted to transfer answers to the official machine-scorable answer sheet.

The official test booklet gives the points assigned to each exam question. Points are assigned for each question depending on the significance, difficulty, and complexity of the question.

Test-Taking Strategy

The Special Civil Engineering Surveying Examination is a difficult test and requires thorough preparation in all areas, including multiple-choice test-taking techniques. Easy and difficult questions, with variable point values, are spread throughout the exam. Besides contending with the nature and difficulty of the exam itself, many examinees spend too much time on difficult problems and leave insufficient time to answer the easy ones. You should avoid this. The following system can be very beneficial to you.

step 1: Work on the easy questions immediately and record your answers on your official machine-scorable answer sheet.

step 2: Work on questions that require minimal calculations and record your answers on your official machine-scorable answer sheet.

step 3: When you get to a question that looks "impossible" to answer, mark a "?" next to it on your official test booklet, and go ahead and guess. Mark your "guess" answer on your answer sheet and continue.

step 4: When you face a question that seems difficult but solvable, mark an "X" next to it on your official test booklet; it may require considerable time in searching for relevant information in your books, references, or notes. Continue to the next question.

step 5: When you come to a question that is solvable but you know requires lengthy calculations or is time-consuming, mark a "+" sign next to the question on your official test booklet. For this question, you know exactly where to look for relevant information in your books, references, or notes.

Based on the number of exam questions and allotted time, on the average, you should not spend more than 2.5 minutes per question. Thus, a lengthy or "time-consuming" question is one that will take you more than 2 minutes and 30 seconds to answer. Don't spend time deciding whether a question should receive a "+" or a "X" because you will waste valuable time. Therefore, work fast and confidently as the intent of this system is save you precious time.

After you have gone over the entire exam, your official test booklet clearly shows the questions that are already answered and questions that still require your attention.

Next:

step 1: The best and most successful approach is to go back and tackle: (a) the "+" questions, (b) the "X" questions, and (c) the "?" questions. As

you proceed, eliminate your pluses, crosses, and question marks.

step 2: Recheck your work for careless mistakes.

step 3: Set aside the last few minutes of your exam period to mark answer spaces for all the unanswered questions on your official machine-scorable answer sheet. There is no penalty for guessing. Only questions answered correctly will be counted toward your score.

How To Use This Book

Civil Surveying Practice Exam for the California Special Civil Engineer Examination works with surveying textbooks. This section tells you how to effectively use the practice exam. It is to your advantage to take all the recommendations of this section into consideration before taking the sample practice exam.

You should take the sample practice exam only after you have reviewed a wide range of textbooks in this field and grasped the subject materials in depth. Optimize your exam preparation by becoming familiar with the content and topics of the examination. Finally, it is suggested that you solve as many sample problems as possible and take some multiple-choice test-taking strategies into consideration prior to taking the sample practice exam.

You will benefit by simulating the official exam constraints and conditions, following the examination administration rules and directions exactly, and giving yourself two and a half hours to answer the problems. A timer works fine. If friends or associates are taking this exam, you may want to get together for a group exam simulation. When it comes to scoring and going over the provided solutions, a group discussion will help you to understand the subject material more thoroughly.

For taking the practice exam, follow the steps listed as follows.

step 1: Use the answer sheet provided to record your answers. Keep in mind that answers written on the exam itself will not be scored. Do not look at the solutions until after you have completed the sample exam.

step 2: Set a timer for two and one-half hours, and begin working.

step 3: When the time is up, stop working on the problems. If you finish sooner, go to the next step. However, it is a good practice to recheck your work if you have time. The rechecking habit could serve you very well on the official exam.

step 4: To score your exam, refer to the instructions in "How To Score Your Exam." Use the answer key to determine your score. This step will enable you to tell what areas you may need to review for a better grasp of the subject matter.

step 5: Interpret your score using the instructions in "How to Interpret Your Score" to appraise your performance.

step 6: Study the provided solutions and review your textbooks and references for those areas where you either missed or guessed. You may have correctly answered questions on certain areas but feel a need to understand their concepts more in depth; study these areas as well.

How To Score Your Exam

For the Special Civil Engineering Surveying Examination, the examination test plan contains the content areas of the exam with the percentages assigned to each defined content area, totaling 100%. The percentages assigned to each content area are the approximate proportion of total test points; however, the test plan does not reveal the total test points in advance. This makes it difficult to anticipate the exact number of problems for each test plan area of the exam. Note that the total test points for this examination varies from exam to exam.

The exam in *Civil Surveying Practice Exam for the California Special Civil Engineer Examination* adopts a total score of 264 for its 52 multiple-choice problems. The point values for each problem are given next to the problem statements as an aid to the examinees. (The point values appear on the actual exam as well.) In this book, these point values are also printed on the answer keys.

On the official exam, after initial scoring, any problem that the Board finds to have a content problem may be deleted. In the event of deletion, the point value of the deleted problem becomes zero and the total number of points possible on that exam is adjusted accordingly.

You will face the official exam with a higher probability of success by going through the scoring process on your practice exam as well as taking the exam. The scoring process will give you an idea of what is needed for you to overcome your weaknesses in order to successfully pass the Special Civil Engineering Surveying Examination.

How To Interpret Your Score

The following table lists statistics for the Special Civil Engineering Surveying Examination that have been obtained from the Board.

	percent passing	cutscore
April 1998	33	184 of 300
October 1997	44	160 of 300
April 1997	43	166 of 300

In this table, the "cutscore" column is very important. Based on the given statistics, if you score above 60% of the total examination score, you have a chance of passing. The probability of success on the actual exam will rise when higher scores are obtained on these practice exams.

Again, this book is devised to help you in appraising areas of your strengths and weaknesses during your exam preparation. This sample practice exam-taking approach enables you to concentrate your review on those subject areas where you have weaknesses. As a result, you can more effectively prepare yourself for the actual exam.

Practice Exam
Problems

Practice Exam
Answer Sheet

1.	Ⓐ	Ⓑ	Ⓒ	Ⓓ		31.	Ⓐ	Ⓑ	Ⓒ	Ⓓ
2.	Ⓐ	Ⓑ	Ⓒ	Ⓓ		32.	Ⓐ	Ⓑ	Ⓒ	Ⓓ
3.	Ⓐ	Ⓑ	Ⓒ	Ⓓ		33.	Ⓐ	Ⓑ	Ⓒ	Ⓓ
4.	Ⓐ	Ⓑ	Ⓒ	Ⓓ		34.	Ⓐ	Ⓑ	Ⓒ	Ⓓ
5.	Ⓐ	Ⓑ	Ⓒ	Ⓓ		35.	Ⓐ	Ⓑ	Ⓒ	Ⓓ
6.	Ⓐ	Ⓑ	Ⓒ	Ⓓ		36.	Ⓐ	Ⓑ	Ⓒ	Ⓓ
7.	Ⓐ	Ⓑ	Ⓒ	Ⓓ		37.	Ⓐ	Ⓑ	Ⓒ	Ⓓ
8.	Ⓐ	Ⓑ	Ⓒ	Ⓓ		38.	Ⓐ	Ⓑ	Ⓒ	Ⓓ
9.	Ⓐ	Ⓑ	Ⓒ	Ⓓ		39.	Ⓐ	Ⓑ	Ⓒ	Ⓓ
10.	Ⓐ	Ⓑ	Ⓒ	Ⓓ		40.	Ⓐ	Ⓑ	Ⓒ	Ⓓ
11.	Ⓐ	Ⓑ	Ⓒ	Ⓓ		41.	Ⓐ	Ⓑ	Ⓒ	Ⓓ
12.	Ⓐ	Ⓑ	Ⓒ	Ⓓ		42.	Ⓐ	Ⓑ	Ⓒ	Ⓓ
13.	Ⓐ	Ⓑ	Ⓒ	Ⓓ		43.	Ⓐ	Ⓑ	Ⓒ	Ⓓ
14.	Ⓐ	Ⓑ	Ⓒ	Ⓓ		44.	Ⓐ	Ⓑ	Ⓒ	Ⓓ
15.	Ⓐ	Ⓑ	Ⓒ	Ⓓ		45.	Ⓐ	Ⓑ	Ⓒ	Ⓓ
16.	Ⓐ	Ⓑ	Ⓒ	Ⓓ		46.	Ⓐ	Ⓑ	Ⓒ	Ⓓ
17.	Ⓐ	Ⓑ	Ⓒ	Ⓓ		47.	Ⓐ	Ⓑ	Ⓒ	Ⓓ
18.	Ⓐ	Ⓑ	Ⓒ	Ⓓ		48.	Ⓐ	Ⓑ	Ⓒ	Ⓓ
19.	Ⓐ	Ⓑ	Ⓒ	Ⓓ		49.	Ⓐ	Ⓑ	Ⓒ	Ⓓ
20.	Ⓐ	Ⓑ	Ⓒ	Ⓓ		50.	Ⓐ	Ⓑ	Ⓒ	Ⓓ
21.	Ⓐ	Ⓑ	Ⓒ	Ⓓ		51.	Ⓐ	Ⓑ	Ⓒ	Ⓓ
22.	Ⓐ	Ⓑ	Ⓒ	Ⓓ		52.	Ⓐ	Ⓑ	Ⓒ	Ⓓ
23.	Ⓐ	Ⓑ	Ⓒ	Ⓓ						
24.	Ⓐ	Ⓑ	Ⓒ	Ⓓ						
25.	Ⓐ	Ⓑ	Ⓒ	Ⓓ						
26.	Ⓐ	Ⓑ	Ⓒ	Ⓓ						
27.	Ⓐ	Ⓑ	Ⓒ	Ⓓ						
28.	Ⓐ	Ⓑ	Ⓒ	Ⓓ						
29.	Ⓐ	Ⓑ	Ⓒ	Ⓓ						
30.	Ⓐ	Ⓑ	Ⓒ	Ⓓ						

1 *(6 points)*

What is the area of a sector with a central angle of 47° within a curve that has a radius of 1100.00 ft?

- (A) 10.34 ac
- (B) 10.76 ac
- (C) 10.99 ac
- (D) 11.39 ac

2 *(4 points)*

Which of the following best describes an area of 1 ac?

- (A) a rectangle 150 ft by 290 ft
- (B) a rectangle 2 ch by 5 ch
- (C) a square 203.71 ft on each side
- (D) a circle with a radius of 116.78 ft

3 *(6 points)*

The field measurements at the point of beginning on a closed traverse are N 4000.000 and E 7000.000; the calculated coordinates at the same point are N 4000.482 and E 6999.727. What is the linear error of closure of the traverse?

- (A) 0.273 ft
- (B) 0.427 ft
- (C) 0.554 ft
- (D) 0.782 ft

4 *(4 points)*

What is the theoretically correct value for the sum of the exterior angles for a closed traverse of 8 courses?

- (A) 180°
- (B) 360°
- (C) 1600°
- (D) 1800°

5 *(4 points)*

Which of the following professional services can a civil engineer legally solicit in the state of California?

- (A) boundary surveys
- (B) construction staking
- (C) infrastructure design
- (D) all of the above

6 *(4 points)*

Which of the following lengths is equivalent to half a chain?

- (A) 37 ft
- (B) 50 links
- (C) 66 links
- (D) 80 links

7 *(6 points)*

Which of the following parameters varies with the scale factor in a Lambert projection as used in a geodetic survey?

- (A) the standard parallel
- (B) the mapping angle
- (C) the latitude
- (D) the longitude

Refer to this graphic for Probs. 8 through 13.

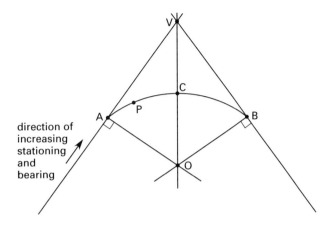

8 *(6 points)*

If the center angle is 56° and the radius is 5000.00 ft, what is the shortest approximate distance between point A and point B?

(A) 590 ft

(B) 660 ft

(C) 2700 ft

(D) 4700 ft

9 *(4 points)*

How many parameters of a horizontal curve must be known in order to calculate the remaining parameters?

(A) 1

(B) 2

(C) 3

(D) 4

10 *(4 points)*

If the tangent distance is 2658.55 ft and the length of the arc is 4886.92 ft, what is the approximate distance from point O to point V?

(A) 4790 ft

(B) 5660 ft

(C) 5730 ft

(D) 6300 ft

11 *(4 points)*

The station line is which of the following?

(A) from point A to point V to point B

(B) from point A to point C to point B

(C) from point A to point O to point B

(D) determined by the design engineer

12 *(4 points)*

If the degree of curvature (arc definition) is 1.1459°, what is the radius?

(A) 3200 ft

(B) 3800 ft

(C) 4200 ft

(D) 5000 ft

13 *(6 points)*

What is the perpendicular distance from point P to a point on the back tangent that is 100 ft from the begin curve point? The radius is 5000 ft.

(A) 1.0 ft

(B) 2.5 ft

(C) 3.0 ft

(D) 9.5 ft

14 *(4 points)*

Which of the following is true of contour lines as shown on a topographic map?

(A) A single contour line can split into two.

(B) Two single contour lines can join into one.

(C) A single contour line can cross with another single contour line.

(D) A single contour line can end within a map.

15 *(8 points)*

An aerial photograph was taken using a camera with a focal length of 6 in. The plane was flying at an altitude of 4000 ft above mean sea level, and the shot was taken of a point on the ground whose elevation was 400 ft above mean sea level. What is the scale of the photograph?

(A) 1 in:10 ft

(B) 1 in:50 ft

(C) 1 in:600 ft

(D) 1 ft:10 ft

16 *(4 points)*

When setting construction stakes, the guard stake (or guard marker) is usually set behind the reference stake (or ginney). The guard stake is marked with notation indicating cut or fill. What do the terms cut and fill refer to in this situation?

(A) the earthwork required as measured from the original ground to the design finished grade

(B) the difference in elevation between the ginney and the design finished grade

(C) the earthwork required as measured from the original ground to the design subgrade elevation

(D) all of the above

17 *(4 points)*

What is the purpose of a turning point in a bench circuit?

(A) It serves as a point of known elevation immediately after the instrument has been repositioned.

(B) It serves as the point about which the error of closure is distributed.

(C) It serves as a temporary benchmark (TBM).

(D) It serves as the point upon which an instrument is to be repositioned.

18 *(6 points)*

A line of magnetic bearing of N 12°32′ E and magnetic declination of 5° E is observed today. What was its magnetic bearing 100 years ago if the magnetic declination at that time was 2° E?

(A) N 12°32′ E

(B) N 14°32′ E

(C) N 15°52′ E

(D) N 19°32′ E

19 *(4 points)*

Which of the following federal provisions created the first surveys of public lands?

(A) the National Lands Act

(B) the Land Ordinance Act

(C) the Public Lands System Act

(D) the System of National Boundaries Act

Refer to this graphic, based on the U.S. System of Rectangular Surveys, for Probs. 20 through 24.

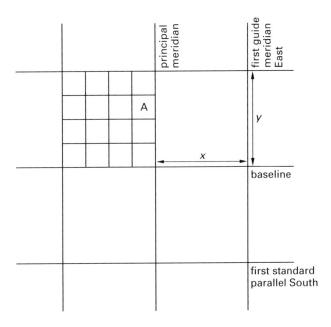

20 *(4 points)*

What is the distance labeled x?
- (A)   5000 ft
- (B)   10,000 ft
- (C)   5 mi
- (D)   24 mi

21 *(4 points)*

What is the distance labeled y?
- (A)   5000 ft
- (B)   10,000 ft
- (C)   5 mi
- (D)   24 mi

22 *(4 points)*

What designation is given to the section labeled A?
- (A)   township 3 North
- (B)   range 1 West
- (C)   township 3 North, range 1 West
- (D)   township 1 West, range 3 North

23 *(4 points)*

What is the area of the section labeled A?
- (A)   6 mi by 6 mi
- (B)   100 ch by 100 ch
- (C)   80 ch by 80 ch
- (D)   8000 ft by 8000 ft

24 *(4 points)*

What is the term used for the area bounded by the labels x and y?
- (A)   a township
- (B)   a range
- (C)   a quadrangle
- (D)   a section

25 *(8 points)*

A curve with a radius of 5000 ft has a central angle of 56°. The distance between the PC and a point on the back tangent (with a station of 152+00.00) is 7821.20 ft. Find the stations of the PC and PT.

(A) 220+24.20 (PC)
     260+05.12 (PT)

(B) 223+21.20 (PC)
     262+07.12 (PT)

(C) 230+21.20 (PC)
     279+08.12 (PT)

(D) 230+21.20 (PC)
     279+09.12 (PT)

26 *(4 points)*

A slope stake is, by definition, set at which point on the proposed cross section of a new facility?

(A) at (or referenced to) the proposed reference line, such as the centerline of a roadway

(B) at the point where the proposed side slope intersects the original ground

(C) at the outermost point from the proposed reference line, such as at the hinge point of a proposed roadway

(D) at any point designated by the design engineer or the field engineer

27 *(4 points)*

What type of survey control does a benchmark provide?

(A) horizontal control only

(B) vertical control only

(C) as designated on the recorded map

(D) both (A) and (B)

28 *(8 points)*

What is the equivalent, in mean solar hours, minutes, and seconds, from Greenwich of a point on the ground at longitude 110°15′32″?

(A) 6°41′2.13″

(B) 7°21′2.13″

(C) 8°3′5.44″

(D) 69°44′28″

29 *(4 points)*

Which definition most closely describes a spiral curve?

(A) a curve with three different radius points

(B) a curve with a constantly increasing or decreasing degree of curvature

(C) a curve that is parabolic on its approach and departure

(D) all of the above

30 *(4 points)*

What is the usual function of a spiral curve in a highway design?

(A) to transition from a tangent to a circular curve

(B) to transition from a parabolic curve to a tangent

(C) to transition from a tangent to a parabolic curve

(D) to avoid an obstruction

Refer to this graphic for Probs. 31 through 36.

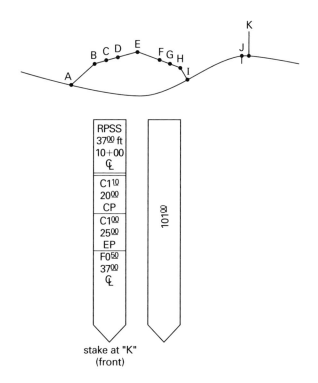

stake at "K"
(front)

31 *(4 points)*

At which point on the cross section would a slope stake be placed?

(A) point E

(B) point K

(C) point A or I

(D) point J

32 *(6 points)*

At which point on the cross section is the reference point slope stake shown?

(A) point E

(B) point K

(C) point A or I

(D) point J

33 *(4 points)*

What is the term used for the stake shown at point K?

(A) the back stake

(B) the forward stake

(C) the reference stake

(D) none of the above

34 *(6 points)*

The first entry under the double strike line on the coded stake indicates which of the following?

(A) that the excavation required at point H is 1.1 ft

(B) that the finished grade elevation at point H is designed to be 1.1 ft below the elevation of the stake set at point J

(C) that the embankment required at point H is 1.1 ft

(D) that the difference in elevation between point I and point H is 1.1 ft

35 *(6 points)*

The second entry under the double strike line on the coded stake indicates which of the following?

(A) that the excavation at point F is 1.0 ft

(B) that the embankment at point F is 1.0 ft

(C) that the difference in elevation between point G and point F is 1.0 ft

(D) that the difference in elevation between the finished grade at point F is 1.0 ft below the elevation of the stake set at point J

36 *(4 points)*

The code on the back of the stake at point K indicates which of the following?

(A) that the elevation of the top of the stake at point J is 101.00 ft

(B) that the offset distance from point K to point E is 101.00 ft

(C) that the design finished grade elevation of point E is 101.00 ft

(D) that the offset distance from point A to point I is 101.00 ft

37 *(4 points)*

Which of the following practices are restricted to licensed land surveyors in the state of California?

(A) establishing, in the field, the centerline of a proposed highway

(B) calculating of the deflection angles required in establishing a traverse used as control in the design and construction of a proposed building

(C) setting property corners in a new residential subdivision

(D) A and C only

38 *(4 points)*

Which of the following instruments incorporates an Electronic Distance Measurement (EDM) device?

(A) a transit

(B) a theodolite

(C) a total station

(D) a level

39 *(4 points)*

What is the theoretically correct value for the sum of the interior angles for a closed traverse of 8 courses?

(A) 876°

(B) 1080°

(C) 1174°

(D) 1440°

40 *(8 points)*

A large parcel of land is to be subdivided such that parcel 1 is 6 ac.

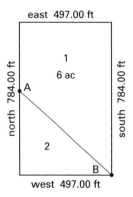

What is the length of boundary AB?

(A) 692.29 ft

(B) 715.95 ft

(C) 738.15 ft

(D) 752.74 ft

Refer to this graphic for Probs. 41 through 46.

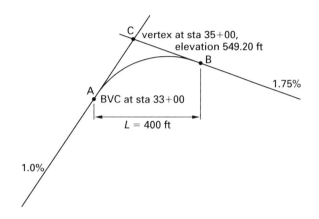

41 *(4 points)*

The length of the curve is defined as which of the following?

(A) the distance along a horizontal datum from point A to point B

(B) the distance along the curve from point A to point B

(C) the distance along the tangents from point A to point C to point B

(D) designated by the design engineer

42 *(8 points)*

At what station does the curve crest?

(A) 33+93.71

(B) 34+35.40

(C) 34+45.40

(D) 35+11.00

43 *(8 points)*

If the vertex is at an elevation of 549.20 ft at sta 35+00.00, what is the approximate elevation of point A?

(A) 546.90 ft

(B) 547.20 ft

(C) 548.10 ft

(D) 548.30 ft

44 *(8 points)*

What is the elevation difference between BVC and a point on the curve at sta 35+00.00?

(A) 0.62 ft

(B) 0.97 ft

(C) 1.24 ft

(D) 32.71 ft

45 *(6 points)*

What is the station at point B?

(A) 36+00.00

(B) 36+25.00

(C) 37+00.00

(D) 37+50.00

46 *(6 points)*

What is the highest elevation on the curve?

(A) 547.93 ft

(B) 548.44 ft

(C) 548.62 ft

(D) 549.38 ft

47 *(4 points)*

Which of the following are used in defining a vertical highway curve?

(A) a circular curve

(B) a spiral curve

(C) a parabolic curve

(D) a broken back curve

48 *(4 points)*

What is a blue top used for in survey staking?

(A) a stake that is driven such that its top is set to design finished grade

(B) a stake that is used to set flowline grades for water lines

(C) a stake that is set at 10 ft intervals on a station offset line

(D) a stake used for controlling the boundaries of a new roadway

49 *(6 points)*

What is the purpose of positioning the instrument such that the foresight and backsight are approximately equal in length when performing a leveling circuit?

(A) Errors due to the line of sight not being perfectly horizontal tend to cancel each other.

(B) Errors due to the line of sight not being perfectly vertical tend to cancel each other.

(C) Errors due to the telescope's varying degree of focus when reading the rod tend to cancel each other.

(D) Both (B) and (C) are true.

Refer to this graphic for Probs. 50 through 52.

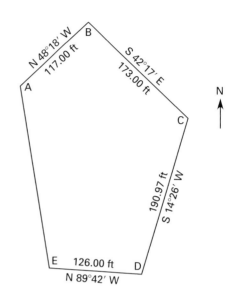

line	bearing	distance	latitude	departure
AB	N 48°18′ W	117.00 ft	77.83′	87.36′
BC	S 42°17′ E	173.00 ft	−127.99′	116.39′
CD	S 24°26′ W	190.97 ft	−184.94′	−47.60′
DE	N 89°42′ W	126.00 ft	0.66′	−126.00′

point	y	x
A	0.0	0.0
B	77.83 ft	87.36 ft
C	−50.16 ft	203.75 ft
D	−235.10 ft	156.15 ft
E	−234.44 ft	30.15 ft
A	0.0	0.0

50 *(4 points)*

What is the double meridian distance, DMD, of course AB?

(A) 81.90 ft

(B) 82.53 ft

(C) 84.23 ft

(D) 87.36 ft

51 *(4 points)*

What is the double meridian distance, DMD, of course BC?

(A) 211.78 ft

(B) 234.55 ft

(C) 256.11 ft

(D) 291.11 ft

52 *(8 points)*

What is the enclosed area of the traverse using the coordinates method?

(A) 0.70 ac

(B) 1.00 ac

(C) 1.60 ac

(D) 2.50 ac

Maximize your chances of passing...

Register this book!

To be fully prepared for your examination, you need the latest information. We would like to keep you up-to-date on our new products as they become available. To receive our announcements, please complete and return this card.

Title of Book

Name _____ Edition _____

Address

Phone

Email Address

To find out about our other products right away, call us toll-free at 800-426-1178 or visit our web site at **www.ppi2pass.com**

Practice Exam
Solutions

Professional Publications, Inc.

Practice Exam
Answer Key

#	Answer	Point Value		#	Answer	Point Value
1.	D	6		31.	C	4
2.	B	4		32.	D	6
3.	C	6		33.	D	4
4.	D	4		34.	B	6
5.	D	4		35.	D	6
6.	B	4		36.	A	4
7.	C	6		37.	C	4
8.	D	6		38.	C	4
9.	B	4		39.	B	4
10.	B	4		40.	B	8
11.	B	4		41.	A	4
12.	D	4		42.	B	8
13.	A	6		43.	B	8
14.	C	4		44.	A	8
15.	C	8		45.	C	6
16.	B	4		46.	A	6
17.	A	4		47.	C	4
18.	C	6		48.	A	4
19.	B	4		49.	A	6
20.	D	4		50.	D	4
21.	D	4		51.	D	4
22.	C	4		52.	B	8
23.	A	4				
24.	C	4				
25.	C	8				
26.	B	4				
27.	B	4				
28.	B	8				
29.	B	4				
30.	A	4				

Total point value = 264

Score: _____

1 *The answer is D.*

$$\frac{A}{\pi r^2} = \frac{I}{360°}$$

$$A = \left(\frac{I}{360°}\right)(\pi r^2) = \left(\frac{47°}{360°}\right)(\pi)(1100 \text{ ft})^2$$

$$= 496{,}284 \text{ ft}^2$$

Convert this area to acres.

$$A = \frac{496{,}284 \text{ ft}^2}{43{,}560 \ \dfrac{\text{ft}^2}{\text{ac}}} = 11.39 \text{ ac}$$

2 *The answer is B.*

Use intuitive trial and error.

$$1 \text{ ch} = 66 \text{ ft}$$
$$(66 \text{ ft})(2) = 132 \text{ ft}$$
$$(66 \text{ ft})(5) = 330 \text{ ft}$$
$$(132 \text{ ft})(330 \text{ ft}) = 43{,}560 \text{ ft}^2 = 1 \text{ ac}$$

An area of 1 ac is best described by a rectangle of 2 ch by 5 ch.

3 *The answer is C.*

The linear error of closure is defined as

$$\sqrt{(\text{difference in N})^2 + (\text{difference in E})^2}$$

$$\text{difference in N} = 4000.482 \text{ ft} - 4000.000 \text{ ft}$$
$$= 0.482 \text{ ft}$$
$$\text{difference in E} = 6999.727 \text{ ft} - 7000.000 \text{ ft}$$
$$= -0.273 \text{ ft}$$

The error is

$$\sqrt{(0.482 \text{ ft})^2 + (-0.273 \text{ ft})^2} = 0.554 \text{ ft}$$

4 *The answer is D.*

$$(n+2)(180°) = (8+2)(180°) = 1800°$$

5 *The answer is D.*

A licensed civil engineer can legally solicit work that his license does not cover, provided the work is ultimately performed under the supervision of a person with the appropriate license. (The codes are B&P 6735, 6735.3, 6735.4, 6738.)

6 *The answer is B.*

There are 100 links in a Gunter's chain. Half of a chain is equal to 50 links.

7 *The answer is C.*

In a Lambert projection, the latitude varies with the scale factor.

8 *The answer is D.*

Compute the long chord.

$$C = 2r \sin\left(\frac{A}{2}\right) = (2)(5000 \text{ ft}) \sin\left(\frac{56°}{2}\right)$$
$$= 4694.72 \text{ ft}$$

9 *The answer is B.*

At least two parameters, such as distances and angles, must be know in order to calculate the remaining parameters of a horizontal curve.

10 *The answer is B.*

$$T = 2658.55 \text{ ft}$$
$$\text{LC} = 4886.92 \text{ ft}$$

Solve for the radius, R_1, and the external distance, E, to compute the distance from point O to point V, which equals $R + E$.

$$T = R \tan\left(\frac{I}{2}\right)$$

$$2658.55 \text{ ft} = R \tan\left(\frac{I}{2}\right)$$

$$R = \frac{2658.55 \text{ ft}}{\tan\left(\dfrac{I}{2}\right)}$$

$$\text{LC} = R(I)$$

$$4886.92 \text{ ft} = R(I)$$

$$\frac{4886.92 \text{ ft}}{I} = R$$

Therefore,

$$\frac{2658.55 \text{ ft}}{\tan\left(\dfrac{I}{2}\right)} = \frac{4886.92 \text{ ft}}{I}$$

$$(2658.55 \text{ ft})I = (4886.92 \text{ ft})\left(\tan\frac{I}{2}\right)$$

By trial and error,

$$I = 0.9774 \text{ rad} = \left(\frac{180°}{\pi \text{ rad}}\right)(0.9774 \text{ rad}) = 56°$$

Therefore,

$$R = \frac{T}{\tan\left(\dfrac{I}{2}\right)} = \frac{2658.55 \text{ ft}}{\tan 28°} = 5000 \text{ ft}$$

$$E = R\left(\tan\frac{I}{2}\right)\left(\tan\frac{I}{4}\right)$$
$$= (5000 \text{ ft})(\tan 28°)(\tan 14°)$$
$$= 662.85 \text{ ft}$$

The distance from point O to point V is

$$5000 \text{ ft} + 662.85 \text{ ft} = 5662.85 \text{ ft}$$

11 *The answer is B.*

The station line is from point A to point C to point B. Stationing on a horizontal curve is along its arc, whereas the stationing on a vertical curve is along a horizontal datum drawn between the BVC and EVC.

12 *The answer is D.*

Using the conversion equation, for D in degrees,

$$D = \frac{5729.6}{R}$$

$$R = \frac{5729.6}{1.1459°} = 5000 \text{ ft}$$

13 *The answer is A.*

Using tangent offsets,

$$x = R \sin\alpha$$
$$100 \text{ ft} = (5000 \text{ ft})(\sin\alpha)$$
$$\sin\alpha = \frac{100 \text{ ft}}{5000 \text{ ft}} = \frac{1}{50}$$
$$\alpha = \arcsin\left(\frac{1}{50}\right) = 1.146°$$
$$y = r(1 - \cos\alpha) = (5000 \text{ ft})(1 - \cos 1.146°)$$
$$= 1.0 \text{ ft}$$

14 *The answer is C.*

A single contour line can cross with another single contour line, for example, in the case of overhanging ledges.

15 *The answer is C.*

$$\text{scale} = \frac{\text{focal length}}{\text{altitude} - \text{point elevation}} = \frac{6 \text{ in}}{4000 \text{ ft} - 400 \text{ ft}}$$

$$= \frac{1 \text{ in}}{600 \text{ ft}}$$

$$= 1 \text{ in:600 ft}$$

16 *The answer is B.*

Cut and fill refers to the difference in elevation between the ginney and the design finished grade.

17 *The answer is A.*

A turning point serves as a point of known elevation immediately after the instrument has been repositioned.

18 *The answer is C.*

The true bearing of the line today is

$$\text{N } 12°32' \text{ E} + 05°00' \text{ E} = \text{N } 17°32' \text{ E}$$

The magnetic bearing of the line 100 years ago was

$$\text{N } 17°32' \text{ E} - 02°00 \text{ E} = \text{N } 15°32' \text{ E}$$

19 *The answer is B.*

The Land Ordinance Act created the first surveys of public lands.

20 *The answer is D.*

A quadrangle is 24 mi by 24 mi square. The distance labeled x is 24 mi.

21 *The answer is D.*

A quadrangle is 24 mi by 24 mi square. The distance labeled y is 24 mi.

22 *The answer is C.*

The U.S. System of Rectangular Surveys would define the area as township 3 North, range 1 West.

23 *The answer is A.*

A township is 6 mi by 6 mi square. The area is 6 mi by 6 mi.

24 *The answer is C.*

The area bounded by the labels x and y is called a quadrangle.

25 *The answer is C.*

The length of the arc is $R\Delta$, where the central angle (Δ) is in units of radians.

$$R\Delta = (5000 \text{ ft})(56°)\left(\frac{\pi}{180°}\right) = 4886.92 \text{ ft}$$

The stationing can be determined by using the point on the back tangent as a beginning point.

location	distance	station
known point		152+00.00
	to PC	78+21.20
PC		230+21.20 (answer)
	arc length	48+86.92
PT		279+08.12 (answer)

26 *The answer is B.*

A slope stake is set at the point where the proposed side slope intersects the original ground.

27 *The answer is B.*

A benchmark provides vertical control only.

28 *The answer is B.*

Greenwich is the prime (or zero) meridian, and each 15° of longitude is one mean solar hour, so

$$\frac{100°15'32''}{15°} = 7°21'2.13''$$

29 *The answer is B.*

A spiral curve is a curve with a constantly increasing or decreasing degree of curvature.

30 *The answer is A.*

The function of a spiral curve is to transition from a tangent to a circular curve.

31 *The answer is C.*

A slope stake would be placed at point A or I.

32 *The answer is D.*

The reference point slope stake is shown at point J.

33 *The answer is D.*

None of the terms are correct. The stake shown at point K is referred to either as the guard stake or the guide marker.

34 *The answer is B.*

The first entry indicates that the finished grade elevation at point H is designed to be 1.1 ft below the elevation of the stake set at point J.

35 *The answer is D.*

The second entry indicates that the difference in elevation between the finished grade at point F is 1.0 ft below the elevation of the stake set at point J.

36 *The answer is A.*

The code indicates that the elevation of the top of the stake at point J is 101 ft.

37 *The answer is C.*

Of the practices listed, setting property corners in a new residential subdivision is the only one restricted to licensed land surveyors in California.

38 *The answer is C.*

The total station incorporates an electronic distance measurement (EDM).

39 *The answer is B.*

$$(N - 2)(180°) = (8 - 2)(180°) = 1080°$$

40 *The answer is B.*

The area of the entire lot is

$$\frac{(784 \text{ ft})(497 \text{ ft})}{43{,}560 \; \dfrac{\text{ft}^2}{\text{ac}}} = 8.94 \text{ ac}$$

Since parcel 1 must contain 6 ac, the area of parcel 2 must be

$$8.94 \text{ ac} - 6.00 \text{ ac} = 2.94 \text{ ac}$$

or

$$2.94 \text{ ac} \left(43{,}560 \; \frac{\text{ft}^2}{\text{ac}} \right) = 128{,}066.4 \text{ ft}^2$$

Parcel 2 is a right triangle with area

$$A = \tfrac{1}{2}bh$$

where b = base and h = height.

$$128{,}066.4 \text{ ft}^2 = \tfrac{1}{2}(497 \text{ ft})h$$
$$h = 515.35 \text{ ft}$$

The hypotenuse $\overline{AB}$ can now be found.

$$(\overline{AB})^2 = (515.35 \text{ ft})^2 + (497 \text{ ft})^2$$
$$\overline{AB} = 715.95 \text{ ft}$$

41 *The answer is A.*

The length of the curve is the distance along a horizontal datum from point A to point B.

42 *The answer is B.*

To find where the curve will crest,

$$x = \frac{g_1}{r}$$
$$g_1 = 1.0\%$$
$$r = \frac{g_2 - g_1}{L} = \frac{-1.75\% - 1.0\%}{4} = -0.66875\%/\text{sta}$$
$$x = \frac{-1.0\%}{\dfrac{-0.6875\%}{\text{sta}}} = 1.454 \text{ sta from BVC}$$

The curve will crest at $(\text{sta } 33{+}00) + (1{+}45.40 \text{ sta}) =$ sta $34{+}45.40$.

43 *The answer is B.*

The elevation can be computed directly using the tangent grade and a portion of the curve length.

$$y = \text{vertex elevation} - g_1 \left(\frac{L}{2} \right)$$
$$= 549.20 \text{ ft} - \left(1 \; \frac{\text{ft}}{\text{sta}} \right) \left(\frac{4 \text{ sta}}{2} \right)$$
$$= 549.20 \text{ ft} - 2.00 \text{ ft}$$
$$= 547.20 \text{ ft}$$

44 *The answer is A.*

Begin by computing the equation of the curve.

$$r = \frac{g_2 - g_1}{L} = \frac{-1.75\% - 1.0\%}{4 \text{ sta}}$$
$$= -0.6875\%/\text{sta} \quad [\text{same as } -0.6875 \text{ ft/sta}^2]$$
$$\Delta_y = \left(\frac{r}{2} \right) x^2 + g_1 x$$
$$= \left(\frac{\dfrac{-0.6875\%}{\text{sta}}}{2} \right) x^2 + (1.0\%)x$$
$$= \left(-0.3438 \; \frac{\text{ft}}{\text{sta}^2} \right) (35 \text{ sta} - 33 \text{ sta})^2$$
$$\qquad + \left(1.0 \; \frac{\text{ft}}{\text{sta}} \right) (35 \text{ sta} - 33 \text{ sta})$$
$$= \left(0.62 \; \frac{\text{ft}}{\text{sta}} \right) (\text{sta})$$
$$= 0.6248 \text{ ft}$$

45 *The answer is C.*

The EVC is a distance $^1/_2 L$ from the vertex.

$$\text{EVC} = (\text{sta } 35{+}00) + \frac{4 \text{ sta}}{2} = \text{sta } 37{+}00$$

46 *The answer is A.*

Applying the equation for the curve,

$$y = \left(\frac{r}{2}\right) x^2 + g_1 x + \mathrm{BVC}_{\text{elevation}}$$

The curve crests at

$$x = \frac{-g_1}{r}$$

$$r = \frac{g_2 - g_1}{L} = \frac{-1.75\% - 1.0\%}{4 \text{ sta}}$$

$$= -0.6875\%/\text{sta} \quad [\text{same as } -0.6875 \text{ ft/sta}^2]$$

$$x = \frac{-1.0\%}{-0.6875\%} = 1.454 \text{ sta from B}$$
$$\quad\quad\quad \overline{\text{sta}}$$

So,

$$y = \left(\frac{-0.6875 \dfrac{\text{ft}}{\text{sta}^2}}{2}\right) (1.454 \text{ sta})^2$$

$$+ \left(1.0 \, \frac{\text{ft}}{\text{sta}}\right) (1.454 \text{ sta}) + 547.20 \text{ ft}$$

$$= 547.93 \text{ ft}$$

47 *The answer is C.*

A parabolic curve is used in defining a vertical highway curve.

48 *The answer is A.*

A blue top is a stake that is driven such that its top is set to design finished grade.

49 *The answer is A.*

The instrument is positioned in that manner because errors due to the line of sight not being perfectly horizontal tend to cancel each other.

50 *The answer is D.*

The DMD of the first course is defined as the departure of the first course.

$$\mathrm{DMD} = 87.36 \text{ ft}$$

51 *The answer is D.*

$$\mathrm{DMD} \, (\mathrm{BC}) = \mathrm{DMD} \, (\mathrm{AB}) + \text{departure} \, (\mathrm{AB})$$
$$+ \text{departure} \, (\mathrm{BC})$$
$$= 87.36 \text{ ft} + 87.36 \text{ ft} + 116.39 \text{ ft}$$
$$= 291.11 \text{ ft}$$

52 *The answer is B.*

Using matrix algebra to solve for the enclosed area, the cross products are

$$\frac{0}{0} \times \frac{87.36}{77.83} \times \frac{203.75}{-50.16} \times \frac{156.15}{-235.10} \times \frac{30.15}{-234.44} \times \frac{0}{0}$$

$$A = \left(\frac{1}{2}\right) \left| \begin{array}{c} \left(\begin{array}{c} (87.36)(-50.16) + (203.75)(-235.10) \\ + (156.12)(-234.44) \end{array} \right) \\ - \left(\begin{array}{c} (203.75)(77.83) + (156.15)(-50.16) \\ + (30.15)(-235.10) \end{array} \right) \end{array} \right|$$

$$= 44{,}914.26 \text{ ft}^2$$

$$A = (44{,}914.26 \text{ ft}^2) \left(\frac{1 \text{ ac}}{43{,}560 \text{ ft}^2}\right) = 1.03 \text{ ac}$$